Mr. P. Naveen Kishore
Dr. A. Kalyan Charan

Development of an Electric Power Weeder for Sustainable Agriculture

AF783656

Mr. P. Naveen Kishore
Dr. A. Kalyan Charan

Development of an Electric Power Weeder for Sustainable Agriculture

A Cost-Effective and Eco-Friendly Alternative to IC Engine-Powered Weeders

LAP LAMBERT Academic Publishing

Imprint
Any brand names and product names mentioned in this book are subject to trademark, brand or patent protection and are trademarks or registered trademarks of their respective holders. The use of brand names, product names, common names, trade names, product descriptions etc. even without a particular marking in this work is in no way to be construed to mean that such names may be regarded as unrestricted in respect of trademark and brand protection legislation and could thus be used by anyone.

Cover image: www.ingimage.com

Publisher:
LAP LAMBERT Academic Publishing
is a trademark of
Dodo Books Indian Ocean Ltd. and OmniScriptum S.R.L publishing group

120 High Road, East Finchley, London, N2 9ED, United Kingdom
Str. Armeneasca 28/1, office 1, Chisinau MD-2012, Republic of Moldova, Europe
Managing Directors: Ieva Konstantinova, Victoria Ursu
info@omniscriptum.com

Printed at: see last page
ISBN: 978-620-8-44019-0

PREFACE

Agriculture remains the backbone of many economies, and efficient farming practices are essential for productivity and sustainability. Weed control plays a crucial role in ensuring optimal crop growth, yet traditional methods such as manual weeding and chemical herbicides present significant challenges. Manual weeding is labor-intensive and time-consuming, while chemical methods pose environmental and health risks. Conventional power weeders, which operate using internal combustion (IC) engines, contribute to pollution, excessive vibrations, and high maintenance costs. Recognizing these drawbacks, this project introduces innovative electric weeders as sustainable and efficient alternatives.

The primary objective of this research is to develop electrically operated power weeders that reduce environmental impact while maintaining high performance and ease of use. These machines replace conventional IC engines with electric motors powered by rechargeable batteries, significantly reducing fuel consumption, emissions, and operational costs. The systems comprise sturdy frames, motors, batteries, controllers, and weeding blades, carefully designed based on the specifications of commercially available power weeders. Additionally, structural analyses, including stress, strain, and deformation assessments, ensure the durability and reliability of these machines in field applications.

As the principal investigators, we embarked on this project with the goal of providing farmers with cost-effective and eco-friendly weeding solutions. By minimizing the dependence on fossil fuels and chemical pesticides, this innovation promotes sustainable farming practices while addressing the high labor demands of traditional weeding methods. The study follows a structured methodology, covering problem identification, literature review, design, fabrication, assembly, and performance testing. The results demonstrate that electric weeders offer viable alternatives to conventional engine-powered models, effectively reducing pollution, vibrations, and maintenance costs while ensuring ease of operation.

This research builds upon previous advancements in agricultural mechanization and electric-powered equipment. The contributions of researchers, engineers, and agricultural experts in the field have provided a strong foundation for this study, and their efforts are sincerely acknowledged.

We extend our heartfelt gratitude to the faculty and staff of Matrusri Engineering College, Hyderabad, for their guidance and support throughout this study. Constructive feedback and suggestions for further improvements are always welcome and will be gratefully acknowledged. The pursuit of sustainable agricultural solutions requires continuous innovation and refinement, and we encourage further research in this field to enhance the efficiency and accessibility of electric weeders for farmers worldwide.

In conclusion, this study aspires to contribute meaningfully to the advancement of sustainable farming practices. Electric weeders present practical, eco-friendly, and cost-effective solutions for weed control in agricultural fields. Their widespread adoption has the potential to reduce reliance on fossil fuels, minimize environmental degradation, and improve farming efficiency.

We sincerely appreciate any suggestions for improvement. Our gratitude extends to LAP LAMBERT Academic Publishing, OmniScriptum GmbH & Co. KG, 120 High Road, East Finchley, London, N2 9ED, United Kingdom, for their support and cooperation throughout the publishing process.

Mr P.Naveen Kishore & Dr. A. Kalyan Charan

Matrusri Engineering College

ABOUT THE AUTHORS

Mr P. Naveen Kishore, a distinguished academician, holds an M.Tech from JNTUH and is currently pursuing a Ph.D. from JNTUK. With 15 years of experience, he serves as an Assistant Professor in the Department of Mechanical Engineering at Matrusri Engineering College, Hyderabad. He has authored 10 research papers, secured three patents, and guided numerous undergraduate projects across diverse disciplines. Actively engaged in professional development, he has both organized and participated in several Faculty Development Programs (FDPs). As an NSS Program Officer, he has spearheaded various social initiatives. His unwavering dedication to education, research, and community service underscores his commitment to academic excellence and societal growth.

Dr. A. Kalyan Charan is an Assistant Professor in the Department of Mechanical Engineering at Matrusri Engineering College, Hyderabad, India. He holds a B.Tech and M.Tech in Mechanical Engineering from JNTU Hyderabad and earned his Ph.D. in Mechanical Engineering, specializing in Heat Transfer, from the same institution. With 15 years of teaching experience, he has been recognized for his excellence in education, receiving the Best Performing Teacher award in 2014, 2015, 2018, and 2020. His research focuses on heat transfer, thermodynamics, and combustion processes, with a strong emphasis on optimizing energy systems. Dr. Charan has authored 25 technical papers, won the Best Technical Paper Presentation award at ICMMSE 2020, and successfully completed five research projects. He is a life member of ISTE, SAE INDIA, IEI, and ISRD and has also published five books.

ACKNOWLEDGEMENT

We sincerely appreciate the esteemed faculty members of the Matrusri Engineering College, Hyderabad, Telangana, for their invaluable support and collaboration in conducting the research presented in this book. Their guidance and expertise played a crucial role in shaping this work into a comprehensive study.

Our heartfelt gratitude extends to the Management of Matrusri Engineering College for their support and permissions, as well as to the Laboratory Staff for their essential contributions to the research process.

We acknowledge the contributions of all those who, directly or indirectly, played a role in the successful completion of this book. We welcome and greatly appreciate suggestions for improvement and error detection from our readers, as their insights will be sincerely valued.

A special note of gratitude goes to our parents, teachers, family, and friends, whose unwavering support and encouragement have been instrumental throughout this journey.

Finally, we extend our sincere thanks to LAP LAMBERT Academic Publishing, OmniScriptum GmbH & Co. KG, London, United Kingdom, for their cooperation and support throughout the publishing process, enabling this book to reach its audience.

Mr P.Naveen Kishore & Dr. A. Kalyan Charan

Matrusri Engineering College

CONTENTS

List of Figures

List of Tables

ABSTRACT

A power weeder is an agricultural machine used to remove unwanted weeds that hinder crop growth. Traditional power weeders operate using internal combustion (IC) engines, which contribute to pollution, generate vibrations, and require high maintenance costs.

To address these drawbacks, an electric weeder has been introduced. This machine operates on battery power, making it a more sustainable alternative. Mechanical weeding methods are preferred due to their numerous advantages over chemical and manual methods. By converting a conventional power weeder into an electrically operated system, the environmental impact is significantly reduced, and maintenance costs are lowered.

The electric weeder utilizes a motor in place of the IC engine, with batteries serving as the primary power source. Its system comprises a frame, motor, battery, controller, and blades. The design is based on the average specifications of commercially available power weeders. Additionally, stress, strain, and deformation analyses are conducted to assess the system's structural integrity and durability.

Overall, the results indicate that the proposed electric weeder is efficient, cost-effective, environmentally friendly, and user-friendly, making it a viable alternative to traditional IC-engine-powered weeders. This innovation effectively mitigates pollution and vibration issues while maintaining high performance and ease of operation.

CHAPTER-I

1. Introduction

1.1 Weed

A weed is any plant that grows in an undesired location, often interfering with agricultural or aesthetic purposes. While certain plants may be beneficial in gardens, farms, or plantations, they become weeds when they negatively impact crop yield, reduce agricultural value, or harm the environment. Some plants are classified as weeds regardless of where they grow due to their invasive nature.

Weeds pose significant challenges in agriculture as they compete with crops for nutrients, water, and sunlight, ultimately reducing farm productivity. They can contaminate produce, overtake pastures, and, in some cases, be hazardous to livestock. Given that agriculture is the backbone of India, effective weed control is essential to maintaining crop quality and preventing the detrimental effects of weed infestations. Removing weeds from fields is a crucial practice to ensure healthier yields and sustainable farming.

Fig.1. 1: Weeds in a field

1.1.1 Characteristics of Weeds

Weeds resemble other plants but possess distinct traits that categorize them as unwanted vegetation. One of their key characteristics is their prolific seed production, particularly among annual weeds. For instance, wild oats (Avena fatua) produce approximately 250 seeds per plant, while wild amaranth (Amaranthus viridis) can generate nearly 11 million seeds per plant. Studies have shown that among 61 perennial weed species, the average seed production capacity is around 26,500 seeds per plant.

Weeds are highly adaptable and can thrive under adverse environmental conditions by adjusting their seed production and growth based on available moisture and temperature. They can germinate even in poor soil-moisture conditions, exhibit rapid growth, and mature faster than most cultivated crops, ensuring early seed production.

Furthermore, weed seeds remain viable for extended periods without losing their ability to germinate. For example, annual meadow grass (Poa annua) and scarlet pimpernel (Anagallis arvensis) can remain viable for about eight years, creeping thistle (Cirsium arvense) for 20 years, and field bindweed Convolvulus arvensis) for approximately 50 years.

Weed seeds also possess remarkable dispersal capabilities, spreading through wind, water, animals, and human activities. In many cases, their seeds closely resemble crop seeds in size and shape, facilitating unintentional transportation along with cultivated seeds. These survival strategies make weeds persistent and challenging to control in agricultural and natural ecosystems.

1.1.2 Effects of Weeds

Weeds significantly consume farmers' time and hinder agricultural development. The process of weeding, while essential, remains labor-intensive. In modern agriculture, there is a growing need for non-chemical weed control methods to ensure food safety. Consumers demand high-quality food products and are increasingly concerned about food safety. Advancements in physical weed control mechanisms may provide a viable solution that aligns with both consumer expectations and environmental sustainability.

Weeds have a substantial impact on agricultural productivity. On average, weed infestations lead to a 5% loss in agricultural production in developed countries, 10% in less developed countries, and up to 25% in the least developed countries. In India, weed-related yield losses exceed those caused by pests and diseases. The extent of yield loss varies depending on the crop, with some crops experiencing over a 50% reduction in yield due to weed competition.

Weeds compete aggressively with crops for essential resources such as water, soil nutrients, light, and space, leading to diminished crop yields. Studies indicate that weeds can deprive crops of approximately 47% nitrogen (N), 42% phosphorus (P), 50% potassium (K), 39% calcium (Ca), and 24% magnesium (Mg). Additionally, weeds serve as alternate hosts for insects, pests, diseases, and various microorganisms, further exacerbating agricultural losses.

Certain weeds release toxic substances into the soil, which can be detrimental to crop plants, human health, and livestock. Some weeds, such as Datura, Argemone, and Brassica, contaminate agricultural produce, making it unsafe for human consumption. The presence of weed seeds in harvested crops can also lead to unpleasant odors, reducing marketability. Moreover, weeds pose a significant fire hazard and create nuisances along railway tracks, roadsides, airport runways, forests, and industrial sites.

Effective weed management strategies are crucial for maintaining agricultural productivity, ensuring food safety, and protecting the environment from the harmful effects of weed infestations.

1.2 Weed Control Methods

In India, weed control is primarily carried out through mechanical and chemical methods. Weed control involves managing weed infestations to ensure profitable crop growth and facilitate other human activities efficiently. Over the years, researchers have experimented with various techniques, achieving varying levels of success.

Effectively controlling weeds requires an understanding of different weed management methods and their strategic application based on specific agricultural needs and environmental conditions. By systematically implementing appropriate weed control measures, farmers can effectively mitigate weed-related issues on their farms.

Broadly, weed control methods can be categorized into five main types, each offering distinct advantages depending on the severity of infestation, crop type, and available resources.

1.2.1 Chemical Weed Control

The use of herbicides has become increasingly popular due to the rising cost and limited availability of manual labor. Chemical weed control methods are rapidly gaining acceptance in the country; however, they also raise significant environmental concerns.

Historically, the use of herbicides for weed control in India has been relatively low. This has primarily been due to the high cost of herbicides, limited awareness of available products, and a lack of understanding of their mechanisms of action. However, with the growing availability of a wide range of herbicides, farmers now have more options to manage different types of weeds in rice cultivation effectively.

Successful chemical weed control requires the careful selection of herbicides and appropriate management practices, which vary depending on the specific rice cultivation system. The following sections present an overview of key herbicides and their effectiveness in weed management.

Fig.1. 2: Weeds in a field

1.2.2 Manual Weed Control

Manual weed control initially involved farmers uprooting weeds by hand. Over time, this method evolved with the introduction of hand tools, progressing from simple sticks to hand-hoes. While manual weeding is highly effective, it demands significant physical effort and energy. The process is labour-intensive and time-consuming, requiring approximately 40 to 60 workers to weed a single hectare of land. Due to high labour costs and the extensive time required, manual weed control has become one of the most expensive aspects of crop production, making it increasingly impractical for large-scale farming.

Table.1. 1: Hand weeding work rates for different types of crops.

Crop	**Hand weeding (h/ha)**
Asparagus	12
Broccoli	50
Carrot	35
Celery	149
Corn	12
Cucumber	74
Dry bean	40
Green bean	30
Green pea	30
Hot pepper	149
Lettuce	94
Mint	45
Onion	158
Peanut	15
Spinach	50
Sweet potato	59
Tomato	92

1.2.3 Mechanical Weed Control

Mechanical weed control methods are widely used, especially in developing countries like India, where agricultural labor is affordable and readily available. These methods are straightforward and easily understood by farmers. The tools and implements used for mechanical weed control are primarily manual or animal-operated.

Mechanical weed control involves various types of weeders, including those operated by human labor, animal-drawn or tractor-drawn weeders, self-propelled weeders, and power weeders. As agriculture became more mechanized, weeding tools initially pulled by draft animals like buffaloes and horses evolved into tractor-adapted implements.

There is a wide variety of mechanical weeders available in the market. These tools function by cutting and uprooting weeds through mechanical tearing, breaking, and cultivation. Typically, mechanical weeding is performed after crops have been planted and emerged. Most mechanical weeders are designed to control weeds between crop rows (inter-row weeding), with only a few specialized machines capable of removing weeds within the crop rows (intra-row weeding).

Among these, inter-row cultivators are the most commonly used machines for mechanical weed control. These implements consist of cultivating tools mounted on a toolbar, which rotate or sweep to displace soil, bury, cut, or uproot weeds efficiently.

Fig.1. 3: Inter-Row weed cultivator

The finger weeder is a straightforward mechanical tool designed for intra-row weeding. It features two sets of steel cone wheels equipped with rubber spikes, known as 'fingers,' which extend outward at a specific angle. Operating with a ground-driven rotary motion, the finger weeder works from the side and beneath the crop row. The rubber fingers penetrate the soil and, just below the surface, effectively uproot small weeds growing in their path.

Fig.1. 4: Finger weeder

Brush weeders feature flexible fiberglass or nylon brushes that rotate around vertical or horizontal axes. These machines primarily uproot weeds while also burying and breaking them. To prevent crop damage, a protective shield or cover can be installed. An operator is responsible for steering the brushes, ensuring they remove as many weeds as possible while avoiding harm to the crop.

Fig.1. 5: Brush weeder

1.2.4 Other Forms Of Weed Control

In addition to conventional approaches, several non-chemical weed control methods utilize alternative energy sources. These include flame weeding, pneumatic weeding, and laser weeding. Each method employs different mechanisms to eliminate weeds without relying on herbicides.

Flame weeding operates by using propane gas to generate intense heat, which raises the temperature of weed plants, either burning their biomass or causing cellular rupture, leading to structural damage and plant death. This method, however, has high energy demands, consuming approximately 28.2 to 131 liters of fuel per hectare (3 to 14 gallons per acre), depending on the application intensity and coverage.

Pneumatic weeding, on the other hand, relies on an air compressor to inject highly pressurized air into the soil, loosening and uprooting small weeds. This method requires a significant power input, typically a 60kW tractor, which is double the energy required for traditional mechanical hoeing.

While these techniques demand substantial energy, they offer viable weed control solutions for organic farming systems. Their non-chemical approach aligns with organic production standards while minimizing soil disturbance, making them environmentally favorable alternatives to conventional herbicides.

Fig.1. 6: A crop-row flame weeder using LPG gas to control weeds inside the crop row.

1.3 Power Weeder

Weeding is essential for crop maintenance, but the use of weedicides is often costly, hazardous, and environmentally harmful. In contrast, mechanical weeding helps maintain soil health by loosening the surface, creating a protective soil mulch that enhances aeration and conserves moisture. Considering these advantages, an engine-operated weeder was designed, developed, and tested for field applications.

The adoption of engine-powered weeders has grown significantly among farmers due to their efficiency and higher output. These machines reduce labor costs and save time, making them a practical alternative to manual weeding. However, they come with certain drawbacks.

One major issue with engine-powered weeders is the high level of vibrations they produce, which can be uncomfortable for the operator. Additionally, these machines require frequent maintenance, adding to operational costs. The rising fuel prices further increase the expense of using engine weeders. Moreover, like conventional vehicles, they emit harmful gases, raising concerns about their environmental impact.

While engine-operated weeders offer significant benefits in terms of efficiency and labor reduction, addressing their drawbacks—such as vibration, maintenance needs, fuel dependency, and emissions—is crucial for sustainable and eco-friendly agricultural practices.

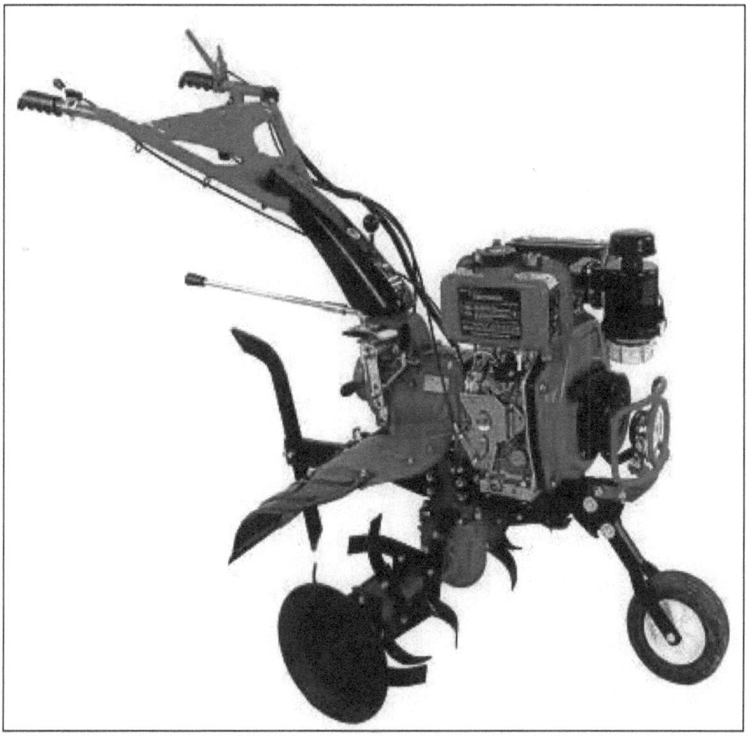

Fig.1. 7: Power Weeder

1.4 Objectives

1. Develop, analyze, and fabricate an Electric Weeder to help farmers efficiently control and remove weeds from their fields with ease.
2. Address the high labor demands of traditional weeding methods, such as cutlasses and hoes, by implementing a mechanical weeding solution suitable for commercial farming.
3. Replace conventional engine-powered weeders with an electric motor, utilizing batteries as the primary power source.

4. Minimize the use of harmful chemical pesticides for weed control, promoting eco-friendly farming practices.
5. Reduce dependence on fossil fuels, leading to cost savings and sustainability.
6. Eliminate pollution and excessive vibrations associated with traditional engine-based weeders, enhancing user comfort and environmental safety.
7. Ensure affordability and accessibility by designing a cost-effective weeding solution.

1.5 Methodology

The method followed to complete the project is as follows

1. Identification
   - Recognizing the problem and defining project objectives.
2. Survey
   - Reviewing relevant literature.
   - Identifying areas for improvement.
3. Design
   - Developing the frame structure.
   - Performing necessary calculations.
4. Evaluation
   - Analyzing the design for feasibility and efficiency.
5. Fabrication
   - Procuring required materials.
   - Assembling and manufacturing components.

6. Assembly

   - Integrating all parts to complete the weeder.

7. Testing

   - Conducting trials to assess the performance of the electric weeder.

8. Results

   - Documenting observations and findings.

9. Report

   - Compiling the final project report.

CHAPTER-II

2. Literature Survey

Development and Evaluation of a Rotary Power Weeder(Authors: J. O. Olaoye and T. Adekanye)

- J. O. Olaoye and T. Adekanye discussed the prevalence of manual weeding in Nigerian agriculture, highlighting its status as the most commonly used weed control method despite being highly labor-intensive.
- Their study emphasized the benefits of mechanical weeders, which can significantly reduce physical strain and improve the working posture of farmers during weeding operations. This, in turn, enhances productivity.
- In response to these challenges, they developed a rotary power weeder. Field performance evaluations of the rotary power weeder demonstrated a field capacity of 0.0712 hectares per hour and a weeding efficiency of 73%. The study underscored the necessity of mechanical weeders in agricultural fields to improve efficiency and ease the burden on farmers.

Design and Fabrication of Power Weeder(Authors: - D. D. Bombe, M. A. Gadakh, S. L. Pathan, S. H. Jain)

- This study explores modifications in the design of a traditional power weeder, aiming to enhance its efficiency in weed removal and cultivation. The redesigned power weeder is specifically engineered to eliminate grass between crop rows while simultaneously facilitating soil cultivation. A key feature of this design is the incorporation of a rotary adjustment mechanism, enabling both weeding and cultivation through a power-operated system.

- The weeder is structured to transmit power to a shaft with blades attached at either end. It is manually driven in a forward direction, with a front-end blade positioned at the roots of the weeds. As the blades rotate, they effectively cut the weeds, ensuring a weed-free cultivation area.
- Despite its effectiveness, the power weeder relies on a diesel engine, which consumes non-renewable fuel resources. Additionally, the setup experiences significant vibrations, which may affect operational stability. This study thoroughly examines both the advantages and disadvantages of power weeders, providing insights into potential improvements for future designs.

Simulation and Analysis of Low-Cost Weeder(Authors: M. Rajashekar, V. K. Heblikar, S. Mohan Kumar)

- This study focuses on the development of a low-cost weeder, emphasizing modeling, simulation, analysis, fabrication, and testing of a three-row weeding machine. The paper explores the concept and essential features of simulation-based design, highlighting its advantages in developing the driving mechanism of the weeding equipment.
- The results demonstrate that the machine would be highly beneficial for small-scale and economically disadvantaged farmers. The functional efficiency of the manually operated weeder was evaluated on different soil types, yielding 81.4% efficiency on loamy soil, 86.5% on clay soil, and 89.4% on sandy soil.

- The primary goal of this project was to design an affordable weeder that could also serve as a cultivator, enhancing its versatility for farmers. Furthermore, design considerations play a crucial role in minimizing manufacturing costs while maintaining efficiency and usability.

Battery assisted four-wheel weeder for reducing drudgery of farmers (Authors: - Mukesh K Singh, S P Singh, M K Singh, Utpal Ekka)

- This study focuses on incorporating batteries in a weeding machine to reduce the physical effort required by farmers.
- A battery-assisted four-wheel weeder was developed using mechatronics concepts for efficient weeding in wide-row crops.
- A new design was introduced to integrate battery and motor usage, featuring a brushed DC motor, battery, power transmission system, drive wheels, and frame.
- The machine also utilizes solar energy for charging the battery, enhancing its sustainability.
- The results indicate that converting a manually operated weeder into a battery-assisted one is a viable and effective solution for wide-row crop weeding.

Design and Fabrication of Manually Operated Double Wheel Weeder (Authors: - G. Selvakumar, S. Dhanasekar, R. Ganesh, S. Mahesh Kumar, S. Giriraj)

- Designed and fabricated a manually operated double-wheel weeder.
- The weeder is human-powered, moving forward as the operator pushes it.

- A blade positioned at the rear end cuts weeds at the roots when the wheels rotate.
- After completing one row, the weeder is shifted to the next, ensuring a weed-free field.
- Key benefits include reduced labor costs, lower time consumption, and cost-effectiveness.
- The simple operation allows easy handling with minimal effort.

Fabrication of Motorized Agriculture Weeder with Grass Collector (Authors: - PVVS Maneendra, S Rajesh, V Shalem Kumar, M Bangaru Raju, K Sai Kumar)

- This paper presents the design and development of a motorized agriculture weeder with a grass collector, originally based on an animal-drawn weeder. A motor was integrated into the design, powered by batteries.
- The system incorporates a modified single-axis motor, converting it into a dual-axis motor for improved functionality.
- It operates through a battery connection, enabling efficient weed removal. A weed collector is added to gather unwanted plants growing between crops.
- The rotary motion of the motor is converted into linear motion using a second-class lever. Sliders are included in the frame to allow horizontal and vertical height adjustments, enhancing operator comfort.
- Traditional weeders rely on engines, which, despite their efficiency, are costly and contribute to environmental pollution. This improved weeder utilizes a DC motor powered by a battery, making it a more eco-friendly

and affordable alternative. Additionally, a grass collector is installed at the bottom to gather weeds and waste materials.

- The study emphasizes the importance of replacing conventional animal-drawn weeders with motorized alternatives.
- The results indicate that the machine is cost-effective, easy to maneuver, and can be operated by any individual in the field. Thanks to the slider adjustments, the weeder can be adapted to the operator's height, ensuring ease of use. Furthermore, it produces less noise compared to other existing machines.

Various aspects of Weeders for Economical Cultivation Authors: -Vivek D. Raut, Prof. B. D. Deshmukh, Prof. Dinesh Dekate)

- This study focuses on analyzing various aspects of weeding equipment for economical cultivation, aiming to reduce labor costs and minimize operator fatigue.
- The primary objective of this review paper is to comprehensively understand the different factors and constraints associated with weeders.
- The paper highlights certain limitations of engine-operated weeding equipment, such as vibration and pollution.
- The performance of weeders can be significantly influenced by factors like adjustable and rotary blades.
- The design considerations of weeding equipment play a crucial role in enhancing its overall efficiency and effectiveness.

CHAPTER-III

3. Problem Specification

3.1 Identification of Problem

1. Weed Removal Importance – Weed removal is a crucial activity in agriculture for maintaining crop health and productivity.
2. Dominance of Chemical Methods – Chemical weed control is more widely used compared to manual and mechanical methods.
3. Environmental Concerns – The adverse effects of chemical herbicides on the environment have led to a shift towards mechanical methods.
4. Toxicity Issues – Herbicides used in chemical weeding are harmful to living organisms and pose toxicity risks.
5. Need for Sustainable Alternatives – There is a growing demand to replace herbicides with more sustainable weed control techniques.
6. Exploration of Innovative Methods – Research is being conducted on alternative physical weed control strategies, including mechanical and thermal methods.
7. Frequent Use of Mechanical Methods – Mechanical weed removal is one of the most commonly used techniques in agricultural fields.
8. Economic Viability – Studies are being carried out to develop cost-effective weed removal methods that do not harm crops.

3.2 Problem Statement

- A power weeder is a conventional machine used in agriculture to remove weeds that obstruct crop growth.

- It typically operates with an internal combustion engine.
- Disadvantages of traditional power weeders:
    - Causes pollution and vibrations.
    - High initial cost and maintenance cost.
- To address these issues, an electric drive battery-powered weeder is proposed.

3.3 Electric Weeder System Overview

- Uses a motor instead of an IC engine, powered by batteries.
- Consists of a frame, motor, battery, controller, and blades.
- Designed based on the average specifications of commercial power weeders.
- Stress, strain, and deformation analyses ensure structural integrity and durability.

3.4 Advantages of the Electric Weeder:

- The transition from internal combustion engines to electric drive systems is a growing trend worldwide.
- Easy to operate.
- Low maintenance.
- Gender flexible (suitable for all users).
- Minimal vibrations.
- No pollution.

CHAPTER IV

4. System Design and Analysis

4.1 Software Used for Design

- SOLIDWORKS CAD is a mechanical design automation software that enables designers to quickly sketch ideas, test features and dimensions, and generate models with detailed drawings.
- The design was created using SOLIDWORKS Edition V2016.
- This user-friendly software includes a comprehensive set of commands and features for efficient design creation.
- It also offers a drafting feature for technical drawings.

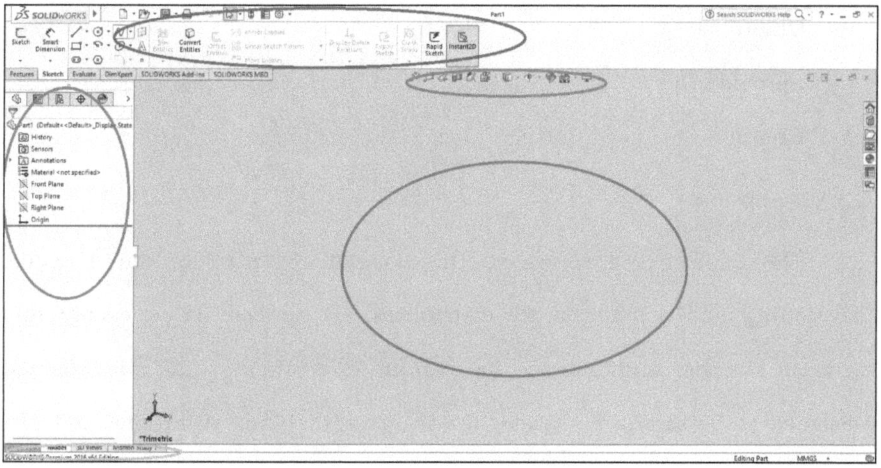

Fig.4. 1: Interface of solidworks

The graphics area (black) serves as the primary space where users can view and interact with a model. At the top of this area, a heads-up view toolbar (green) provides quick access to key functions. On the left side, the feature manager design tree (blue) organizes model components. Below the reference triad, a status bar (purple) displays information at the bottom of the window. The command manager (red) is positioned at the very top, offering essential tools and commands.

4.2 Components List

- Main Frame
- Blades
- Drive Shaft
- Brushless DC Motor
- Wheel
- Lead Acid Battery
- Chain
- Sprocket
- Charger

4.2.1 Mainframe

The main frame serves as the skeletal structure of the machine, functioning as its chassis. It distributes all applied loads evenly and provides support and balance for various components and attachments. Designed to be strong and rigid, the frame withstands vibrations and tilts during operation. Constructed using the arc welding process, it acts as the primary supporting member of the machine. The prime mover is mounted at the front, while the handle is attached at the rear. Additionally, the blade shaft is secured at the center of the base frame.

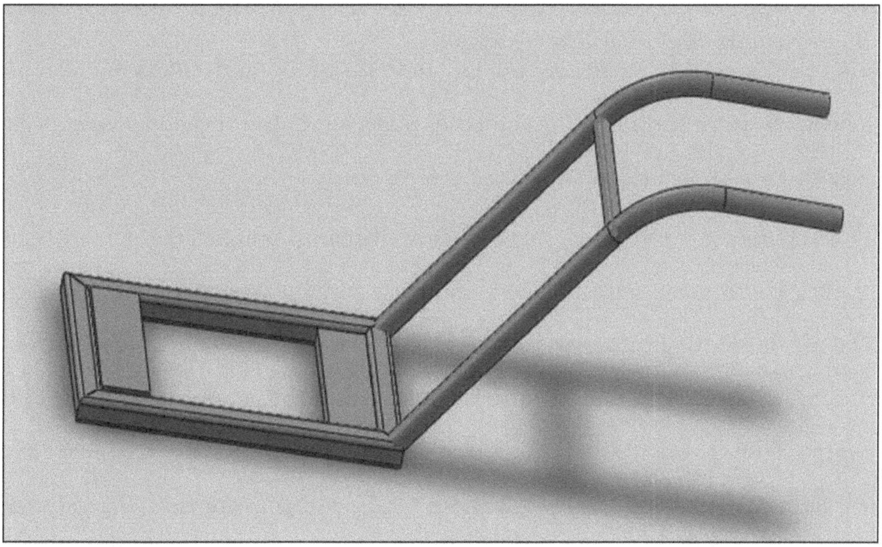

Fig.4. 2: Design of Mainframe

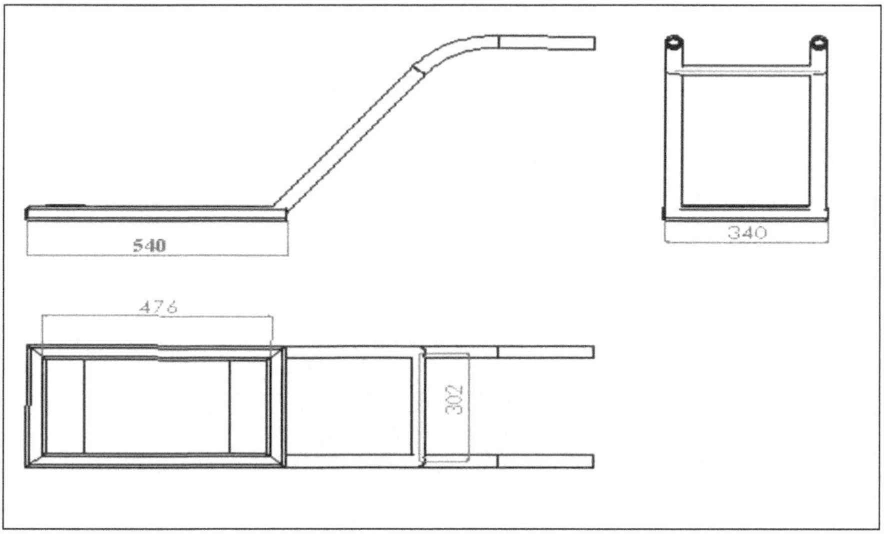

Fig.4. 3: Mainframe dimensions

4.2.2 Blade

A rotavator is an essential machine for soil preparation in agriculture. This motorized equipment utilizes rotating blades to turn and aerate the soil, performing a function similar to cultivators and tillers. While all rotavators serve the same fundamental purpose, different models come with varying capabilities to suit specific farming needs.

By using a rotavator, farmers can efficiently prepare the soil without requiring excessive manual labor, ensuring optimal conditions for planting. The machine's blades play a crucial role in removing unwanted weeds from the soil. These blades, also known as tines, are arranged in sets of three and mounted on a shaft. In total, 16 tines are positioned to maximize soil interaction. Since the blades are the primary components that engage with the soil, they significantly impact weed control and overall soil preparation.

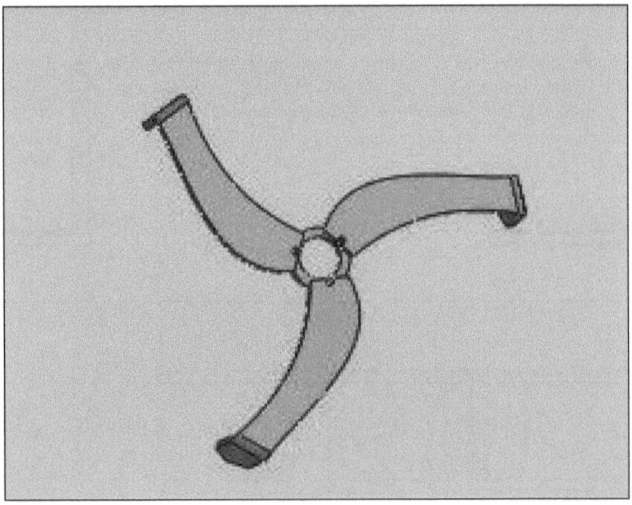

Fig.4. 4: Design of Blade

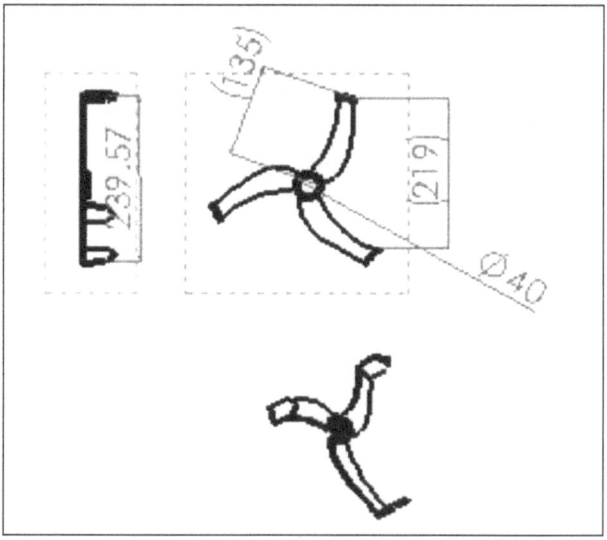

Fig.4. 5: Blade dimensions

4.2.3 Shaft

The shaft, made of mild steel, connects the sprocket and chain system, which transmits power from the motor. Blade sets are attached on both sides of the shaft. It functions as a rotating machine element, facilitating power transmission from the motor via the chain.

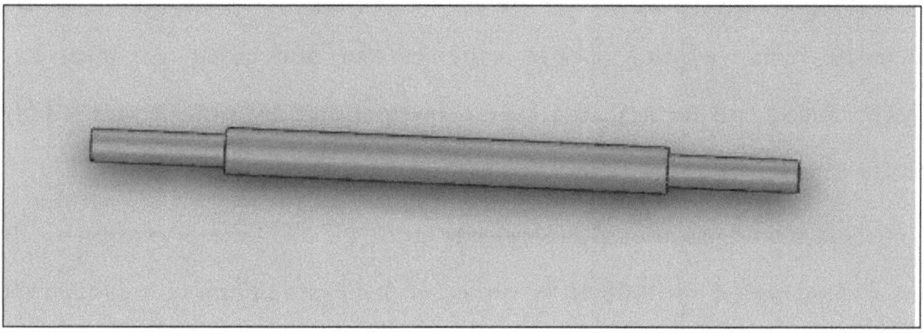

Fig.4. 6: Blade dimensions

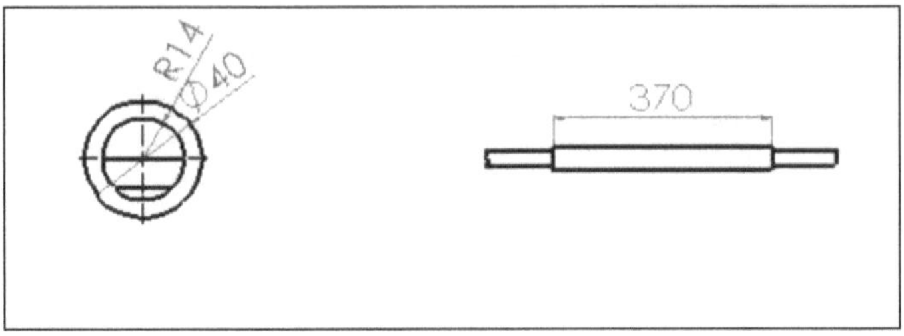

Fig.4. 7: Shaft dimensions

4.2.4 Brushless DC Motor

A DC motor is a type of rotary electrical machine that converts direct current (DC) electrical energy into mechanical energy. The most common types operate using forces generated by magnetic fields. Nearly all DC motors incorporate an internal mechanism—either electromechanical or electronic—to periodically reverse the direction of current flow in specific motor components.

In this system, the motor is mounted at the front of the machine, which is positioned on the mainframe. A Brushless DC (BLDC) motor is used, paired with a controller for precise motion and speed regulation. BLDC motors offer several advantages, including low maintenance, high efficiency, high operating speeds, compact size, and quick response. The motor's speed can be adjusted over a broad range by varying the supply voltage or modifying the current strength in its field windings.

This motor provides the necessary drive for machine displacement. It has a rated speed of 3000 RPM under no-load conditions, a rated power capacity of 750W, and operates with a rated current of 30–45 amps.

Fig.4. 8: Design of motor

Table.4. 1: Motor Specifications

Motor type	**BLDC**
Rated power	750W
Rated speed	3000RPM
Rated torque	2.39N.m
Current	36.76A
Voltage	24V

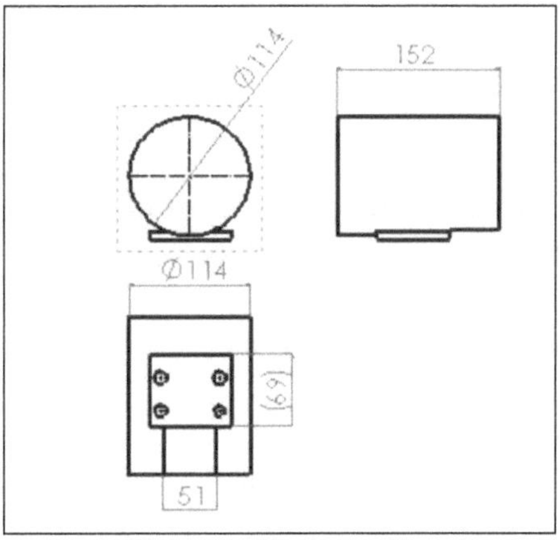

Fig.4. 9: Motor dimensions

4.2.5 Wheel

The wheel is the rotating component of the machine, positioned at the front. A traction wheel is used to improve grip, enhancing the wheel's ability to maintain contact with the ground without slipping.

Fig.4. 10: Wheel

4.2.6 Lead acid Battery

The power source for the machine is a lead-acid battery, which supplies the necessary energy to drive the motor. Lead-acid batteries are cost-effective and offer an efficiency of approximately 80-85%. They also generate minimal heat and come in a wide range of capacities. The system uses a constant 12V battery with a 50AH rating. Two batteries are connected in series to achieve a total voltage of 24V, providing increased power and torque for the motor. These batteries are securely mounted on a tray, which is welded onto the mainframe for stability.

Fig.4. 11: Lead acid battery

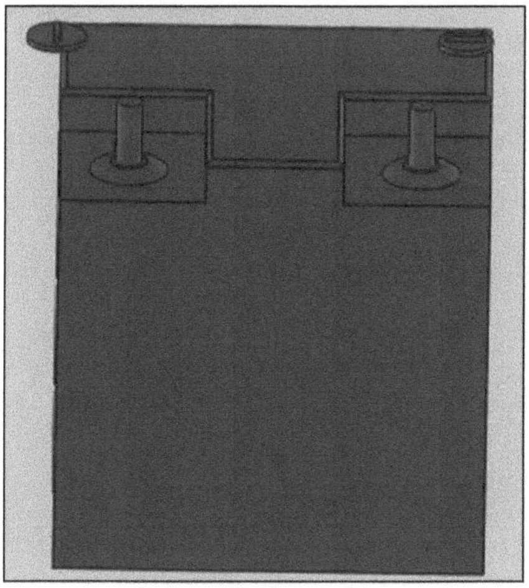

Fig.4. 12: Design of battery

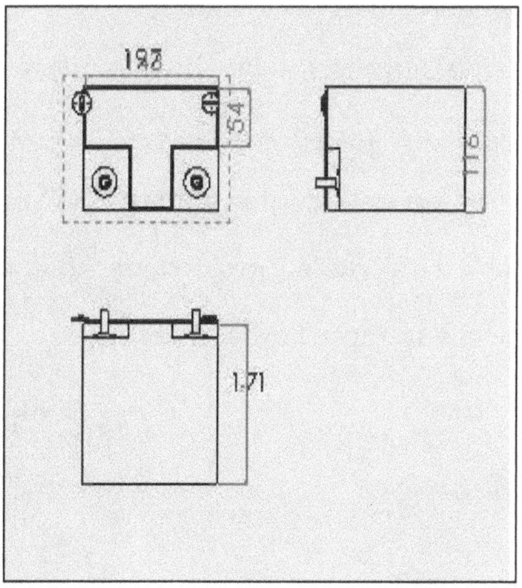

Fig.4. 13: Battery dimensions

4.2.7 Chain

A chain drive is a mechanism for transmitting mechanical power between two points. Commonly, it transfers power to vehicle wheels, especially in bicycles and motorcycles. Beyond vehicles, chain drives are widely used in various machines. In this system, the chain links the motor's shaft to a sprocket, enabling power transmission.

Fig.4. 14: Chain with sprocket

4.2.8 Sprocket

A sprocket, also known as a sprocket-wheel or chain wheel, is a toothed or cogged wheel designed to engage with a chain, track, or other perforated or indented material. The term "sprocket" broadly refers to any wheel with radial projections that interact with a passing chain. It is typically mounted on a shaft.

Fig.4. 15: Sprocket

4.2.9 Charger

The charger is designed to recharge lead-acid batteries using the constant current method, which is both efficient and fast. This charging process prevents battery overheating and helps extend the battery's lifespan.

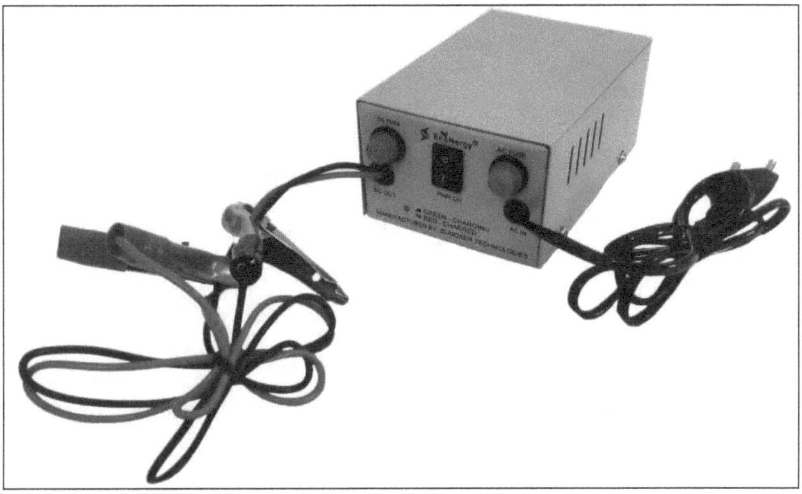

Fig.4. 16: Charger

4.3 Complete Assembly

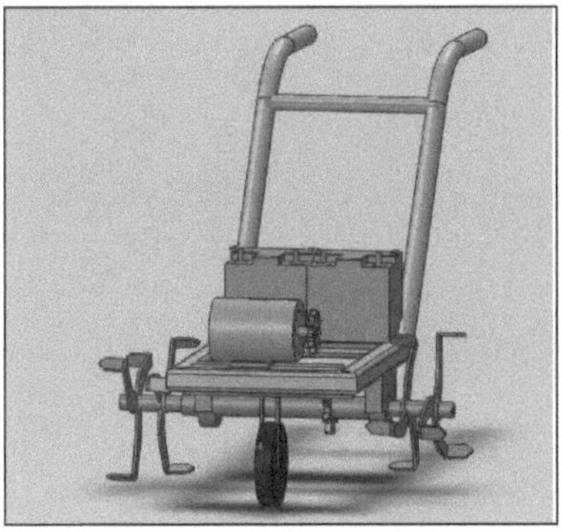

Fig.4. 17: Assembly

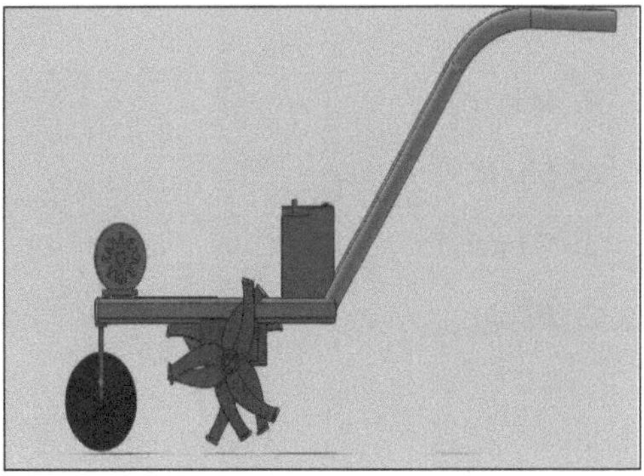

Fig.4. 18: Assembly side view

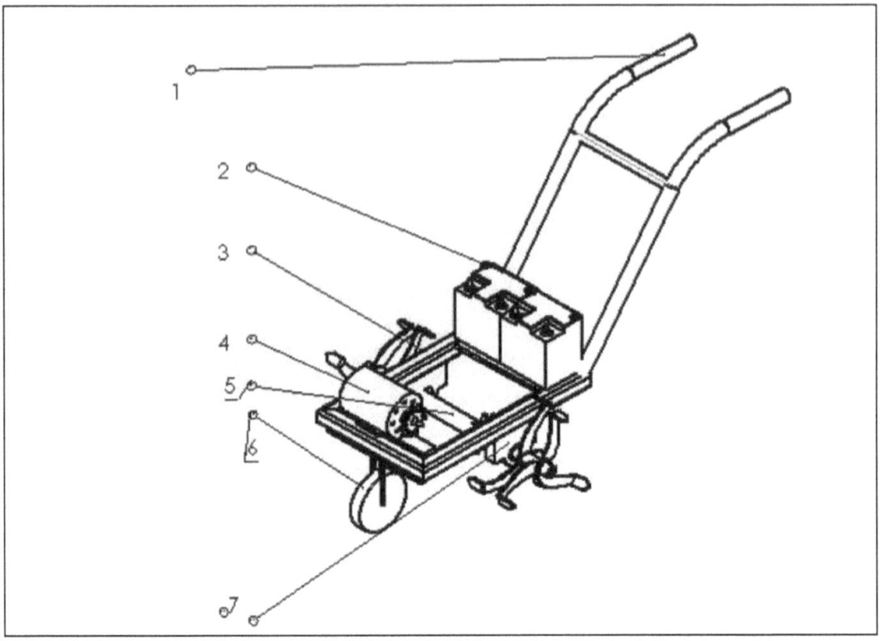

Fig.4. 19: Assembly 3D view

Components

1. Handle
2. Battery
3. Frame
4. Motor
5. Shaft
6. Wheel
7. Blade

4.4 Analysis

4.4.1 Software Used for Simulation

- The simulation was conducted using Ansys Workbench 2016.
- The analysis focused exclusively on mechanical components subjected to weight-bearing conditions.
- Key aspects examined included deformation and shear failure.
- The results indicated that the material possesses a high load-bearing capacity.
- This conclusion is derived from the obtained simulation results.

ANSYS®

Fig.4. 20: Ansys logo

4.4.2 The ANSYS Workbench Interface

The ANSYS Workbench interface primarily includes a Toolbox region, the Project Schematic, the Toolbar, and the Menu Bar. Depending on the analysis type, application, or workspace, additional windows, tables, and charts may also be available. Users can interact with ANSYS Workbench by dragging components or analysis systems from the Toolbox to the Project Schematic or by double-clicking items to initiate default actions.

DesignModeler serves as a geometry editor for existing CAD models and is a parametric, feature-based solid modeler. It enables users to efficiently create 2D sketches, model 3D parts, or import 3D CAD models for engineering analysis preprocessing. The Workbench Simulation module allows users to define environmental loading conditions, solve simulations, and review results in various formats based on the type of simulation.

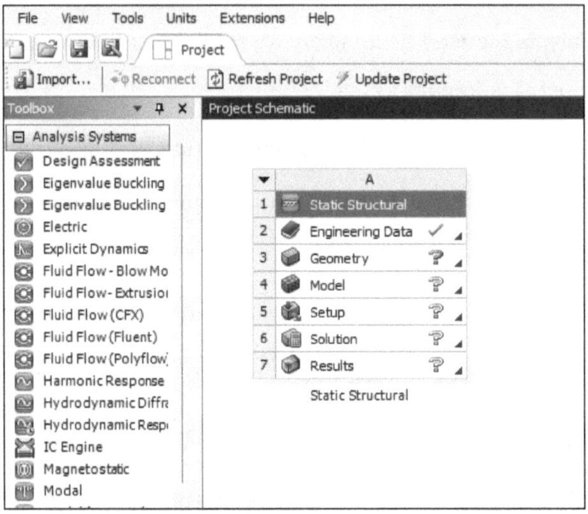

Fig.4. 21: Ansys interface

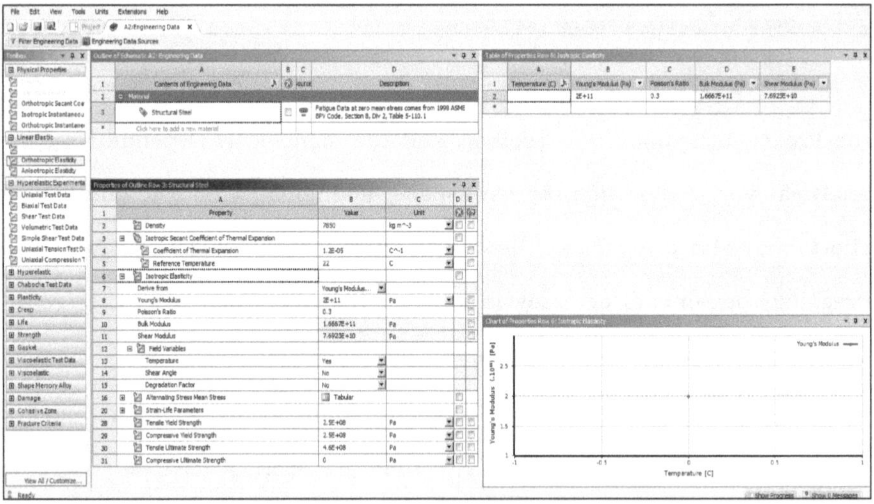

Fig.4. 22: Engineering Data Mode

4.4.3 Material Properties of Mild Steel

Table.4. 2: MS Properties

Properties of Mild Steel	**Values with dimensions**
Density	7.87 g/cc
Tensile Strength, Ultimate	440 MPa
Tensile Strength, Yield	370 MPa
Modulus of Elasticity	205 GPa
Bulk Modulus	140 GPa
Poisson's Ratio	0.29
Machinability	70 %
Shear Modulus	80.0 GPa
Electrical resistivity	0.0000159 Ω-cm

4.5 Analysis of mainframe

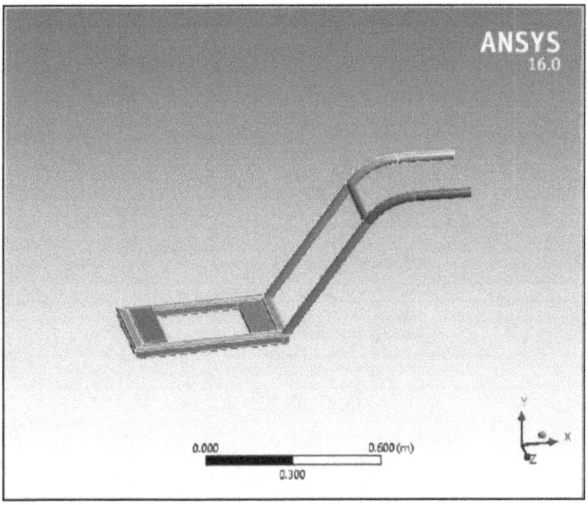

Fig.4. 23: Frame file imported to ansys

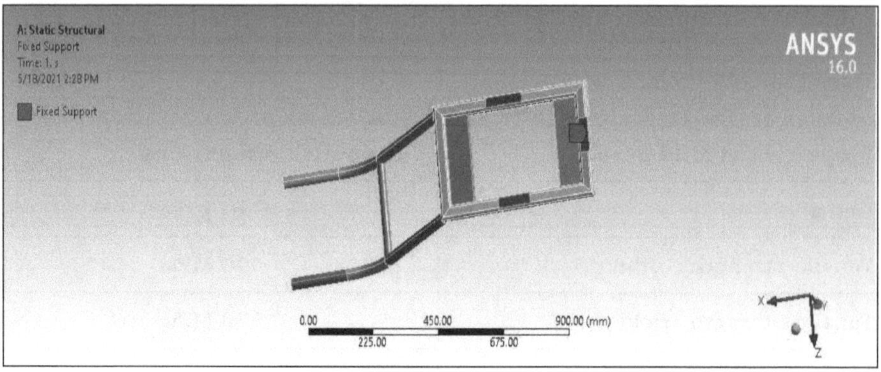

Fig.4. 24: Fixed supports on the frame

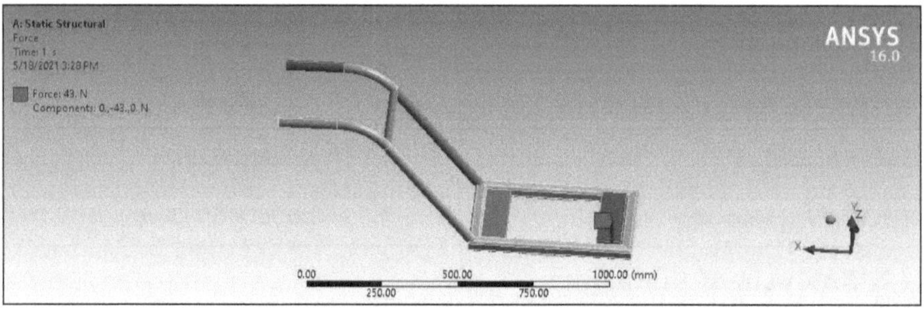

Fig.4. 25: Force 1 Motor

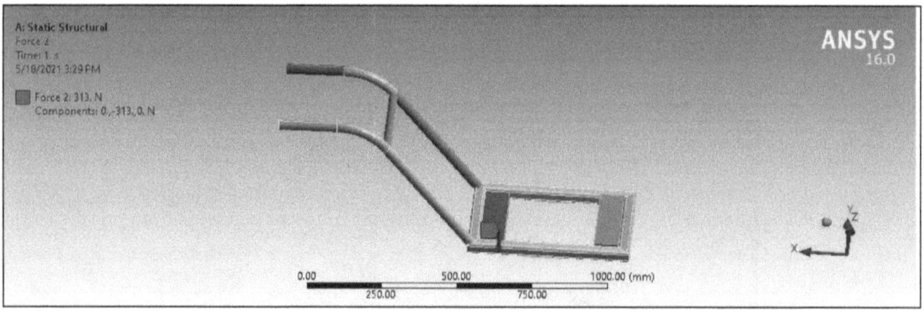

Fig.4. 26: Force 2 Motor

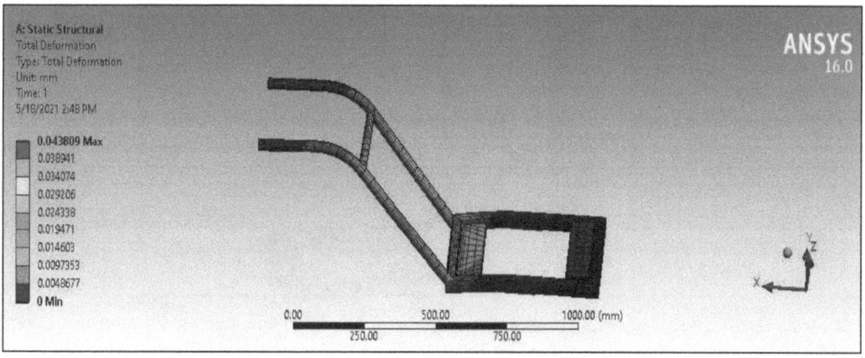

Fig.4. 27: Total deformation with motor and battery weights

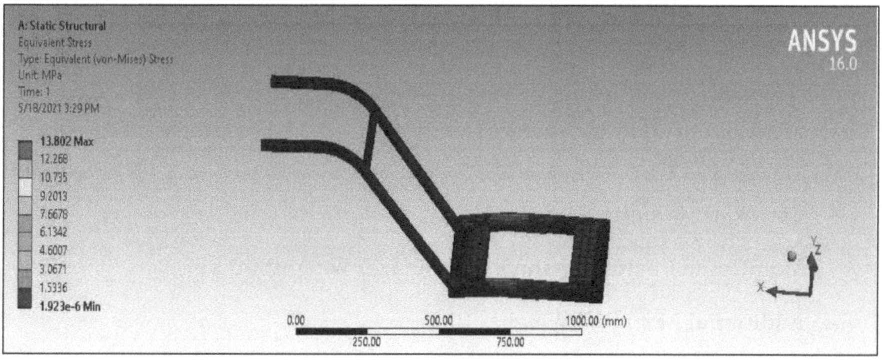

Fig.4. 28: Equivalent stress

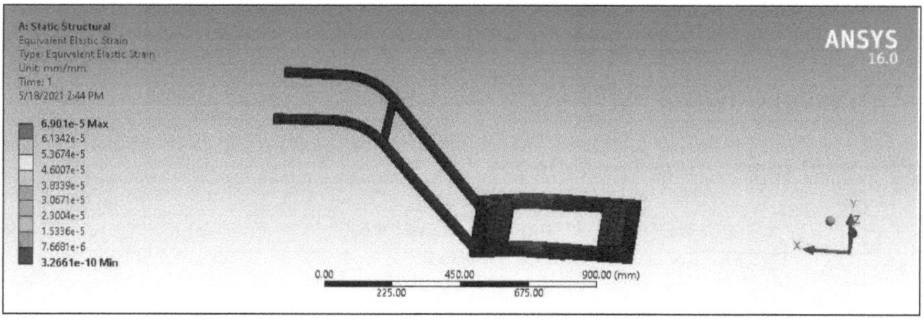

Fig.4. 29: Equivalent strain

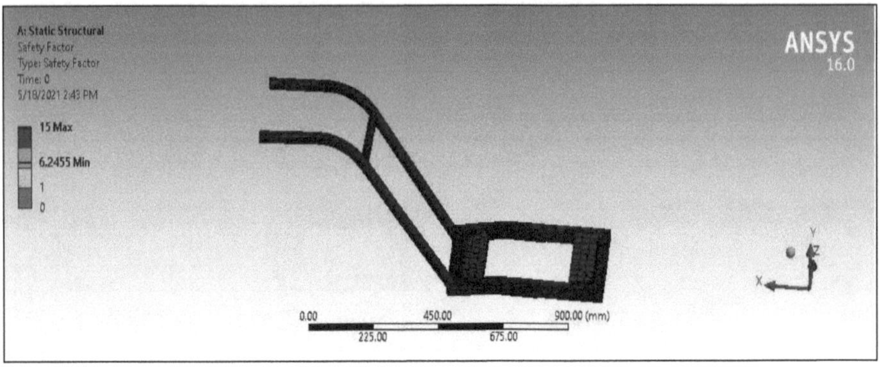

Fig.4. 30: Safety Factor

4.6 Calculations

- Motor Weight: 4.3 kg
- Battery Weight: 16 × 2 = 32 kg
- Controller Weight: 1.25 kg
- Mainframe (including shaft and blades) Weight: 50 kg
- Additional Assumed Weight: 10 kg
- Total Estimated Load: 100 - 120 kg
- Motor Rating Calculation: Load to be pulled = 100 kg
- Required Maximum Speed: 1.2 – 1.5 km/h
- Velocity: 0.64 m/s

Power (P) is calculated as follows:

$$P = \text{Force} \times \text{Velocity}$$

$$= 1000 \times 0.64$$

$$= 640\ \text{N} \cdot \text{m/s} = 640\ \text{watts}$$

- Considering a motor with 750W power and 3000 RPM.

Torque (T) is calculated as follows:

$$\textbf{Torque (T)} = \frac{60\text{P}}{2 * \text{pi} * \text{N}}$$

$$= \frac{60\text{P}}{2 * 3.14 * 3000} = 2.388\ Nm$$

Calculation of Required Battery Backup

- Motor Power (P): 750 W
- Estimated Battery Backup Time: 1.5 hours
- Total Battery Output Voltage: 12V × 2 = 24V
- Required Storage Capacity:

$$= \frac{\text{P} * \text{t}}{\text{V}} = \frac{750 * 1.5}{24} = 47\text{Ah}$$

- Selected Battery Specification: 12V, 50Ah

Battery Specifications:

- Battery Pack Voltage: 24V
- Battery Capacity: 50Ah
- Energy Storage: 1,200Wh (24V × 50Ah)

Charging Time Calculation:

- Voltage: 12V
- Capacity: 50Ah
- Load Current: 10A
- Charging Time:

$$\text{Charging Time} = \frac{\text{Capacity}}{\text{Load Current}} = \frac{50}{10} = 5$$

- Total Charging Time: 300 minutes

CHAPTER V

5. Working and operation

The electric weeder operates on a battery-powered system. Instead of a conventional engine, it utilizes a motor for driving the weeding mechanism. The motor converts the chemical energy stored in the batteries into mechanical energy, which powers the weeding process. Blades are mounted on a shaft, receiving power from the motor for efficient operation. To generate the necessary power, the weeder is equipped with two 12V lead-acid batteries.

5.1 System Operation

1. Start the motor using the on/off switch located on the handle.
2. The machine moves forward with the help of rotating blades.
3. The battery supplies power from the motor to the shaft.
4. Blades are mounted on both sides of the shaft, enabling forward motion when they rotate.
5. The rotating blades cut the weeds effectively.
6. This process allows for efficient weeding with minimal effort and cost.

5.2 Material Specification

Table.5. 1: Material Specification

S.No.	Title	Specification	Quantity No's
1	Brushless DC motor	1 HP, 750kw, 24V, 3000 rpm	1
2	Battery	12V, 50ah	2
3	Charger	24V, 10amps, constant current type	1
4	Transmission	Chain and sprocket	1
5	Frame	Mild steel	-
6	Blades	Mild steel, 12 blades, split into 3 blades on each set, 4 sets	12

CHAPTER VI

6. Conclusions and Future Enhancements

6.1 Observations

The electric weeder has proven to be highly beneficial for farmers, significantly reducing their effort while improving efficiency. Since it operates on a battery, it is portable and easy to handle. This machine is cost-effective and does not require skilled labor for operation, thereby lowering labor costs when used commercially.

As the weeder does not rely on fuel, its maintenance costs are minimal. It requires less time for weeding, eliminating labor wages entirely. Additionally, it reduces both human and animal effort while covering more land compared to conventional methods. The machine operates with low noise and vibration, producing no pollution, making it an environmentally friendly option.

The electric weeder significantly cuts down the overall cost of weed removal operations by completing the task in a shorter time. Its eco-friendly design ensures no health hazards for the operator. The compact structure allows it to remove weeds from complex areas efficiently, reducing the need for frequent field inspections. Furthermore, its design makes it accessible and convenient for use by individuals of any gender.

6.2 Applications

- Used for removing unwanted weeds.
- Serves as an alternative to animal-drawn weeders.
- Suitable for large agricultural fields.
- Effective in fields with a high density of weeds.
- Ideal for regions with labor shortages.
- Applicable for maintaining private lawns.

6.3 Results Analysis

The machine's endurance was tested using Ansys software. Additionally, calculations were performed to determine the required power, torque, battery backup, battery size, and charging time. The endurance test analysis yielded the results presented in the following table.

Table.6. 1: Analysis results

Properties	**Values**	
	Min	**Max**
Total deformation (mm)	0	0.043829
Equivalent stress (N/mm2)	1.923e-6	13.802
Equivalent strain	3.2661e-10	6.901e-5
Safety factor	6.24	15

The results obtained suggest that the frame withstands the given loading conditions. The calculation results and analysis were compared with the existing weeder, confirming that the proposed system meets farmers' budget constraints and reduces labor costs.

6.4 Conclusion

- The primary objective of this project is to replace the traditional engine-powered weeder with an electric weeder, eliminating pollution and reducing vibrations.
- Our proposed design achieves this by utilizing a simple structure where a motor replaces the conventional engine, powered by a battery.
- The electric weeder offers gender flexibility, allowing anyone in the field to operate it with ease. This innovation empowers farmers to manage weeding independently, reducing reliance on laborers.
- By using an electric weeder, zero emissions are ensured, fulfilling the core purpose of developing this machine.

6.5 Limitations

- Requires more time to charge the batteries.
- Heavier compared to traditional weeders.
- Lower efficiency than engine-powered weeders.
- Cannot withstand impact loads.
- Limited usability in certain crop fields due to its width.

6.6 Future Enhancements

The current product is designed for weeding row crops with a row width matching that of the weeder. It operates using electricity, making it an eco-friendly and non-polluting alternative to fuel-based weeders. Additionally, its operating cost is significantly lower compared to fuel-powered options. While petrol or diesel costs approximately 85-90 Rs per liter, electricity costs range from 2.50-6 Rs per unit, depending on the location. A full recharge takes around 6 hours when the battery is completely drained, and the total operating cost is approximately 35-40 Rs for 1.5 hours of operation.

6.6.1 Potential future improvements include:

- Enhancing the weeder's width adjustment for greater adaptability.
- Implementing automation through advanced controllers.
- Integrating multi-tool functionality for broader applications.
- Advancing towards small autonomous machines capable of treating crops and soil based on specific agricultural needs, representing the next stage in modern farming technology.

REFERENCES

[1.] Olaoye, J. O., & Adekanye, T. A. (2012). *Development and Evaluation of a Rotary Power Weeder*. ResearchGate Publications.

[2.] Bombe, D. D., Gadakh, M. A., Pathan, S. L., & Jain, S. H. (2018). *Design and Fabrication of Power Weeder*. 6th International Conference on Recent Trends in Engineering & Technology (ICRTET).

[3.] Rajashekar, M., Mohan Kumar, S., & Heblikar, V. K. (2014). *Simulation and Analysis of Low-Cost Weeder*. International Journal of Research in Engineering and Technology, eISSN: 2319-1163 | pISSN: 2321-7308.

[4.] Singh, M. K., Singh, M. K., Singh, S. P., & Ekka, U. (2019). *Battery-Assisted Four-Wheel Weeder for Reducing Drudgery of Farmers*. Indian Journal of Agricultural Sciences, 89(9), 1434–1438.

[5.] Sai Kumar, K., Bangaru Raju, M., Maneendra, P. V. V. S., Rajesh, S., & Shalem Kumar, V. (2020). *Fabrication of Motorized Agriculture Weeder with Grass Collector*. Journal of Emerging Technologies and Innovative Research (JETIR), JETIR2003141.

[6.] Maheshkumar, M., Ganesh, R., Selvakumar, Dhanasekar, S., & Girira, S. (2017). *Design and Fabrication of Manually Operated Double Wheel Weeder*. International Journal of Innovative Research in Science, Engineering and Technology, ISSN (Online): 2319-8753 | ISSN (Print): 2347-6710.

[7.] Tharoon, T. (2017). *Design and Fabrication of Rotary Tiller Blade*. International Research Journal of Engineering and Technology (IRJET), Volume 04, Issue 1, e-ISSN: 2395-0056 | p-ISSN: 2395-0072.

Kolibri 360 GmbH
Ferdinand-Jühlke-Straße 7
99095 Erfurt, Deutschland
produktsicherheit@kolibri360.de

Druck:
CPI Druckdienstleistungen GmbH
im Auftrag der
Kolibri 360 GmbH
Ferdinand-Jühlke-Str. 7
99095 Erfurt